Editor's Comment

It is December Already! How did we arrive here? Thank God, we did it- hurray!

This last edition for the year 2016 takes a dive back into the issues that hits the nerves of the energy, oil and gas industries. This retrospective outlook is informative, educative and a must read for all stakeholders. I will say path yourself in the back, take a deep breath and be happy that you survived. Girl oh! Boy what a year of incredible challenges but, because as an industry we have been through this before and have learned from our past. That is why we are here today and thriving.

2017 looks promising for the industry- in this edition you will read about New Pipeline Infrastructure Projects that will Increase Natural Gas Takeaway Capacity In 2017. A comprehensive lists of pipeline projects at various stages that have been approved to come onstream in 2017.

Innovative technology that are the catalyst of growth- is moving the industry forward and keeping it relevant to the world. For example, The First Microchannel Gas-To-Liquid Plants is set to Convert Stranded Natural Gas to Marketable Products.

In the meantime, challenges still remain as can be seen in the international Maritime Organization IMO move to tighten Fuel Sulfur Limits on Maritime Transportation Will Spark Changes by Both Refiners and Vessel Operators.

Finally-The largest infrastructure conference – international pipeline, oil and gas safety conference will take place in Houston Texas – March 14-16, 2017. Visit the event site for more details- www.oilandgassafetyconference.com

Happy Thanksgiving and Merry Christmas

- Gloria Towolawi

Contents

USA Oil and Gas Monitor
A RGT Media Communications Corp.

Editor-in-Chief
Gloria Towolawi

Europe Bureau
Esther Coker

Nigeria Bureau
David Arhavbarien

Contributing Editor
Gloria Instead

Reporter
Caleb Motinwo

Advert & Marketing
Jewel Spring
T: 832-486-0095
E: advertise@usaoilandgasmonitor.com

Distribution & sales
Richard Godfirst

Subscribers Service
E: subscribe@usaoilandgasmonitor.com

RGT Media Communications Corp.
Publishers of
USA Oil and Gas Monitor
Workplace Weekly News
GlobalPRPlus

USA Oil and Gas Monitor is published 12 times a year monthly by RGT Media Communications Corp. 10777 Westheimer
#1100

Houston, Texas 77042
Subscription price is $144 per year.
Digital copy $9.99 per download.

December 2016 • Issue 12

Flashback! Hits and Misses in Global Energy Scenario 2016- How the industry fared?

As the year is coming to a close- USA Oil and Gas Monitor- December Edition is taken a look back on the issues that drove the industry forward and the once that were missed opportunities for the energy industry. Fasten your seat belt- as we drive back 11 months earlier from here. These analyses also show USA Oil and Gas Monitor in-depth coverage of the energy industry in one year.

This is a historic victory for the oil industry when the U.S. House of Representatives and the Senate passed the spending bill to fund the federal government through September 2016, which includes a provision lifting the 40-year-old ban on the export of crude oil.

The lifting the 40-year-old ban on the export of crude oil

This is a historic victory for the Global energy industry in which, the United States and over 200 other countries supported an international climate deal aimed at reducing emissions from oil and other fossil fuels, which was largely opposed by the industry.

Paris, France

The signing of the international climate deal

On January 16, 2016, the International Atomic Energy Agency confirmed that Iran had fulfilled its initial nuclear-related obligations under the Joint Comprehensive Plan of Action JCPOA that had been reached in July between the United States, the other P5+1 countries, the European Union, and Iran. As a result, on that day, Implementation Day, the UN, the United States, and the European Union lifted their nuclear-related sanctions against Iran. For European companies, this essentially means that, provided they comply with the EU's export licensing requirements for shipments to Iran, they should be able to engage in most types of transactions with that country.

The Lifting of Sanctions Against Iran

With the growing consensus around global market that the current oil price is not right, stakeholder at the just concluded World Economic Forum 2016, agreed with one voice and express their willing to cut oil production to stabilize the oil market. Missed opportunity- No agreement was reached regarding this issues in 2016.

Oil Market Prices Not Right; OPEC and Non-OPEC Willing to Cut

Partly as a result of a strong domestic production growth, both domestic natural gas consumption and exports of natural gas by pipeline have increased, and exports of liquefied natural gas LNG from the United State this year.

Several new pipeline projects have come online to move natural gas either to nearby market areas in the Mid-Atlantic area New York, New Jersey, and Pennsylvania or to feed into existing infrastructure that delivers natural gas to more distant regions, especially the U.S. Gulf Coast.

New Pipeline Projects

Shell and BG Shareholders Vote in Favor of the Recommended Combination between Shell and BG

Ben van Beurden, CEO of Shell, said: "I am delighted with the positive shareholder vote and the confidence that shareholders have shown in the strategic logic of the combination of Shell and BG.

GE's acquisition of Alstom. Mr. Immelt noted that there are a finite number of attractive acquisition targets, and this opportunity had already been analyzed. Alstom complements GE's capabilities and increases its reach into emerging markets.

Golar and Schlumberger announced the creation of OneLNGSM, a joint venture to rapidly develop low cost gas reserves to LNG. Golar and Schlumberger have 51/49 ownership of the joint venture. OneLNG will be the exclusive vehicle for all projects that involve the conversion of natural gas to LNG, which require both Schlumberger Production Management services and Golar's FLNG expertise.

Successful Merger and Acquisition

December 2016 • Issue 12

While positive market sentiments continue to arise from the output freeze plan being considered by major crude exporters. Including support from expected decreasing US production, higher refinery runs, declining production in several other regions and an increase in unplanned outages. In addition to the Healthy physical oil markets –particularly in Asia amid positive margins – and ongoing strategic stockpiles in China and the U.S. Missed opportunity- No agreement was reached regarding this issues in 2016.

OPEC Output Freeze Plan

The arrival of MOL Benefactor, which is also the largest ship ever to call the Port of Savannah, ushers in a new era of larger vessels and services that will increase capacity, volumes and economic opportunities for Georgia's savannah Port. The Savannah Harbor Expansion Project SHEP will deepen the inner harbor to 47 feet and the outer harbor to 49 feet at mean low water. The outer portion of the harbor is now 15 percent complete with work progressing daily.

Supply Chain- Port of Savannah sees Opportunities with its Panama Canal Expansion Project

This growing demand is in response to growing domestic hydrocarbon gas liquids HGL supply and favorable petrochemical feedstock prices in the United States relative to the international market.

Rightly put: Mark Lashier, Chevron Phillips Chemical's Executive Vice President said, "Demand for the chemicals and plastics business is strong, compelling us to continue to search the globe for the next major investment."

The petrochemical industry uses hydrocarbon feedstocks such as ethane and propane to create plastics, fibers, resins, and a wide range of other consumer and industrial materials.

Most of the ethylene cracker projects currently being developed will not come online until 2017 or 2018, including six large-scale projects announced in 2011 and 2012. Four projects from Dow, ExxonMobil, Chevron Philips, and OxyChem/Mexichem are already under construction, and two projects from Formosa and Sasol have received permitting approval and commitments from investors. Together with capacity expansions at existing facilities, these six new facilities are expected to increase U.S. ethylene production by 40 per cent, to a total of more than 37 million metric tons mt, more than one-fifth of current global ethylene production capacity approximately 150 million mt.

Strong Global Demand for Petrochemicals

The future of global natural gas expansion is supported by strong supply growth, particularly because of unconventional gas and LNG. Despite, the growing role of natural gas as a bridge fuel towards a long term increasingly renewable-based, efficient and sustainable energy system. Natural gas is not fully being utilized by the energy sector as fuel for today which helps to meet the emission target. Hence, a missed opportunity for the industry in 2016. Nonetheless, Natural gas will increase its relative share in the global primary energy supply from 21.4 per cent in 2013 to 23.9 per cent in 2035.

Natural Gas Fuel and Emission Target

The Federal Aviation Administration Reauthorization Law allows for unmanned aircraft system UAS utilization for oil and gas facilities, refineries, pipeline inspection, and response activities. Drone technology will complement the comprehensive safety practices that the industry has in place to ensure that all Americans continue to enjoy the affordable, reliable fuels they depend on. The ability to use drones will allow the industry to use the latest technologies to continue to effectively monitor infrastructure and facilities while minimizing the risk to personnel.

Energy Infrastructure Safety

The oil industry is now most of the way through the largest terms of trade shock in history, since second half of 2014. The good news is that the industry will emerging stronger, much stronger, out on the other side. Last year 3 per cent, real GDP growth 300,000 jobs created the highest number since 2006 before the GFC.

Now rather than abandoning mines or plugging wells, the industry should adapt to lower prices by reducing costs, pushing the frontiers of innovation and increasing market share. And one of the greatest examples of innovation, resilience and expansion is LNG.

Chevron's Gorgon project, Shell's Prelude Floating LNG project are some of the Mega LNG project that have come onstream. In particular, Gorgon is a triumph and has witness a ROI with its first LNG shipment to Japan.

Industry Survive the Cyclical Shock and will Emerge Stronger. Resilience and Innovation seen through the expansion of LNG

December 2016 • Issue 12

EPA announced comprehensive steps to address methane emissions from both new and existing sources in the oil and gas sector. For new, modified and reconstructed sources, EPA finalized set of standards that will reduce methane, volatile organic compounds VOCs and toxic air emissions in the oil and natural gas industry.

These sources include hydraulically fractured oil wells, some of which can contain a large amount of gas along with oil, and equipment used across the industry that was not regulated in the 2012 rules.

EPA updated a number of aspects in the final rule that increase climate benefits, including removing an exemption for low production wells and requiring leak monitoring surveys twice as often at compressor stations, which have the potential for significant emissions. The final rule also provides companies a pathway to align the final standards with comparable state-specific requirements they may have.

EPA Releases Regulatory Standards to Cut Methane Emissions in the Oil and Gas Sector

Incorporate the latest industry standards that establish minimum baseline requirements for the design, manufacture, repair, and maintenance of blowout preventers BOP. The regulation requires an annual review of the repair and maintenance records of the BOP equipment by a BSEE approved third party to ensure that the equipment continues to meet the original design criteria.

Operators are required to use BSEE Approved Verification Organizations BAVOs no later than 1 year from the date when BSEE publishes the list of BAVOs etc.

BSEE Well Control Rule and Timeline

It's easy to blame the industry's current malaise on the sunken oil price, which since 2014 has been way below its 10-year average. But frankly, much of what afflicts the industry is of its own doing – and it predates 2014 by more than 10 years.

Over the many years that operators have been working together, both inside and outside respective companies –; Mr. Harry Brekelmans of Shell said, "our behavior have allowed cost, risk and inefficiency to spread unchecked across the industry's supply chain. This is a missed opportunity that has overlooked for far too long which has come back to hurt the industry.

Low Oil Price Environment Reveals some "dirty little secrets" of the industry

USA Oil and Gas Monitor representing the United States Energy publication for the oil and gas industry and other international journalist from the United Kingdom, Australia, India and Nigeria; at the request of CORE PA Global-tour some of Pennsylvania's innovative companies and industrial suppliers involved in the mining, oil and gas field machinery manufacturing industries.

The purpose- CORE PA Global is an initiative established to increase the visibility of a 53-county footprint of Pennsylvania, to international and domestic investors. CORE PA partners with over 50 economic development organizations and the Commonwealth of Pennsylvania to attract, retain, and grow business and industry in the region.

Special Report on the Oil and Gas, Mining Industries in Rural Pennsylvania

First quarter 2016 financial results from U.S. onshore producers reveal an improving balance between capital expenditure and operating cash flow. Although operating cash flow was the lowest in any quarter in the past five years, larger reductions to capital expenditure brought these companies closest to self-finance when capital investment can be paid for entirely from operating cash flow.

Shale oil and gas shows sign of recovery in first quarter 2016

Energy Infrastructures is the bedrock and key to ensuring global supply of oil and gas. The 'who is who' in the pipeline industry was looked at; the 13 largest pipeline operators in North America which delivers oil and gas to end users were ranked by system capacity.

Thirteen Largest U.S. Interstate Natural Gas Pipeline Operators Ranked by System Capacity

Crude output rose by 400 kb/d in June to an eight-year high of 33.21 mb/d, including newly re-joined Gabon. Saudi Arabia ramped up to a near-record rate of 10.45 mb/d and Nigerian flows partially recovered. Middle East producers sustained record pumping rates, consolidating market share and pushing OPEC's total output 510 kb/d above one year ago.

Robust European demand supported second quarter 2016 global demand growth at around 1.4 mb/d year-on-year, momentum that will be roughly matched through the year as a whole. A modest deceleration is foreseen in 2017, as growth eases to 1.3 mb/d taking average deliveries up to 97.4 mb/d.

Nigeria and Saudi Arabia Consolidated OPEC Market Share

December 2016 • Issue 12

U.S. and China announced six new Eco-Partnership Among the new partnerships is a collaboration between Alabama-based Chemical and Metal Technologies CMT and China's CPI Yuanda Environmental Protection Engineering Company. CPI Yuanda is a subsidiary of State Power Investment Corporation SPIC, one of China's largest generating power companies.

The CMT-CPI Yuanda partnership will evaluate CMT's new cleanup technology for removing heavy metals from industrial wastewater produced during capture of sulfur dioxide in a post-combustion flue gas desulphurization FGD unit. The technology will be tested at SPIC's Hechuan Flue Gas Comprehensive Experimental Base's 2 x 300 MW coal-fired power plant in Chongqing, China. A successful demonstration of CMT's technology for wastewater cleanup would be followed by a demonstration of the technology for flue gas treatment.

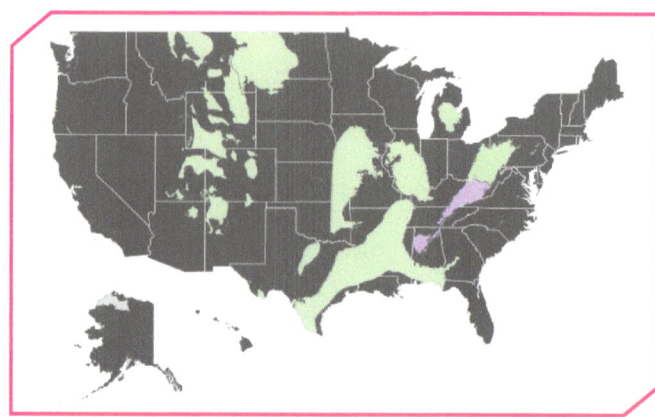

U.S.-China Eco-Partnership to Make Coal Power Plants Cleaner

The southern California hub is the top gateway for trade between the United States and Asia. It supports over a million jobs nationally and generates billions of dollars in economic activity each year.

Lehigh Valley is a two-county region in eastern Pennsylvania. Located one hour north of Philadelphia and west of New York City. Port Newark/Elizabeth Marine Terminal NJ/NY Port of Philadelphia.

DALLAS-FORT WORTH, HOUSTON INTERCONTINENTAL, PORT OF HOUSTON, TEXAS CITY, CORPUS CHRISTI, BROWNSVILLE, HIDALGOMCALLEN, LAREDO, EAGLE PASS and EL PASO. These ports aid quick logistics and distribution decision- making Texas the third largest distribution hub in North America.

Toronto benefits from Canada's strong trade relationships, including the historic North American Free Trade Agreement NAFTA, which created the largest free trade zone in the world, and shares proximity and time zone with the world's largest concentration of economic activity - the northeastern United States. European and Asian markets also offer tremendous trade opportunities.

Intermodal yard is the main reason Chicago remains one of the key transportation hubs in North America.

Top Five North America's Distribution Hubs

Of the 141 operable refineries in the United States, there were 30 with capacity over 100,000 barrels per Atmospheric Crude Oil Distillation.

The 30 Largest U.S. Refineries with Capacity Over 100,000 Barrels per Atmospheric Crude Oil Distillation

Oil and Gas to Remain Dominant-Renewable Energies and Nuclear Expected to Increase Their Share. According to HE Mohammad Sanusi Barkindo, OPEC Secretary General, the future of energy is a future laden with challenges and uncertainties, but also opportunities, however, one thing stakeholders agree on is the fact that the world will need more energy in the decades to come. It is easy to appreciate why.

All of the three main primary sources of energy – oil, gas and coal – will still supply more than three-quarters of the energy mix by 2040. Oil will be at just over 25 per cent, with coal slightly less, and gas slightly more. However, these renewable energies and nuclear are expected to increase their share in the energy mix from 18 per cent in 2015 to 22 per cent by 2040.

The Future of Energy

The regions are Bakken, Eagle Ford, Haynesville, Marcellus, Niobrara, Permian, and Utica

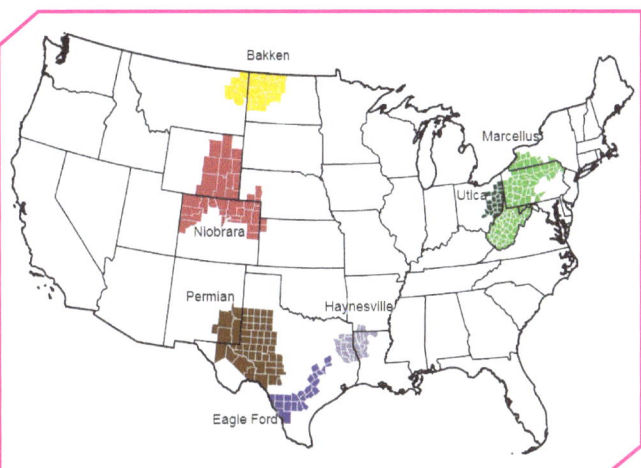

Seven Regions in U. S. Accounts For 92 Per Cent of Domestic Oil and Gas Production

The first U.S. ethane export terminal, located in Marcus Hook, Pennsylvania, about 20 miles southwest of Philadelphia, has an export capacity of 35,000 barrels per day b/d and began shipping ethane cargos in March 2016. Ineos Olefns and Polymers Europe, with ethylene cracker operations in Scotland and Norway, and its partner Evergas, a company specializing in seaborne petrochemical and liquid gas transportation, took delivery of the first ship in a planned eight-vessel fleet of Large Gas Carriers LGC.

Ineos Olefns and Polymers Europe receives U.S. first ethane shipment

December 2016 • Issue 12

Sustained and significant increases in U.S. crude oil exports, however, likely require more than lower shipping costs and sporadic purchases. It would require increased U.S. crude oil production and a significantly wider Brent-WTI price spread.

U.S. crude Oil now exports to 16 countries and Canada, however challenges remains

The company estimates hydrocarbons in place on its acreage position are 75 trillion cubic feet Tcf of rich gas more than 1,300 British Thermal Units and 3 billion barrels of oil in the Barnett and Woodford formations alone.

Apache Corporation Discovers Significant New Resource Play in Southern Delaware Basin

Mexico Largest Importer of U. S. Natural Gas by Pipeline. Argentina Largest Consumer of U. S. LNG Export by Vessel. Mexico No 1 Destination for U. S. Natural Gas Export by Truck. Canada Top Destination for U. S. Natural Gas Export by CNG.

U.S Natural Gas Export by Pipeline, LNG Vessel, Truck and CNG- Top Countries and Volumes, Half Year Analysis

Economic activity and population drive increases in energy use; energy intensity E/GDP improvements moderate this trend. Non-OECD nations drive the increase in total energy use. Increases to non-OPEC oil supplies outside the United States are primarily from Brazil, Russia, Canada, and Kazakhstan. Shale gas, tight gas, and coalbed methane will become increasingly important to gas supplies, not only for the U.S., but also China and Canada.

Global Energy Outlook -At A Glance in 24 Charts

Nine are Hydroelectric Facilities. Four of the world's ten largest power plants are located in China. The Kashiwazaki-Kariwa nuclear power plant in Japan is the largest nuclear plant in the world and the sixth-largest power plant of any type in the world.

The Ten World's Largest Power Plants

20 Billion Barrels of Oil in Texas' Wolfcamp Shale Formation Uncovered

Using a geology-based assessment methodology, the U.S. Geological Survey assessed technically recoverable mean resources of 20 billion barrels of oil and 16 trillion cubic feet of gas in the Wolfcamp shale in the Midland Basin part of the Permian Basin Province, Texas.

This is the largest estimate of continuous oil that USGS has ever assessed in the United States.

The Wolfcamp shale in the Midland Basin portion of Texas' Permian Basin province contains an estimated mean of 20 billion barrels of oil, 16 trillion cubic feet of associated natural gas, and 1.6 billion barrels of natural gas liquids, according to an assessment by the U.S. Geological Survey. This estimate is for continuous unconventional oil, and consists of undiscovered, technically recoverable resources.

The estimate of continuous oil in the Midland Basin Wolfcamp shale assessment is nearly three times larger than that of the 2013 USGS Bakken-Three Forks resource assessment, making this the largest estimated continuous oil accumulation that USGS has assessed in the United States to date.

"The fact that this is the largest assessment of continuous oil we have ever done just goes to show that, even in areas that have produced billions of barrels of oil, there is still the potential to find billions more," said Walter Guidroz, program coordinator for the USGS Energy Resources Program. "Changes in technology and industry practices can have significant effects on what resources are technically recoverable, and that's why we continue to perform resource assessments throughout the United States and the world."

Although the USGS has assessed oil and gas resources in the Permian Basin province, this is the first assessment of continuous resources in the Wolfcamp shale in the Midland Basin portion of the Permian.

Since the 1980s, the Wolfcamp shale in the Midland Basin has been part of the "Wolfberry" play that

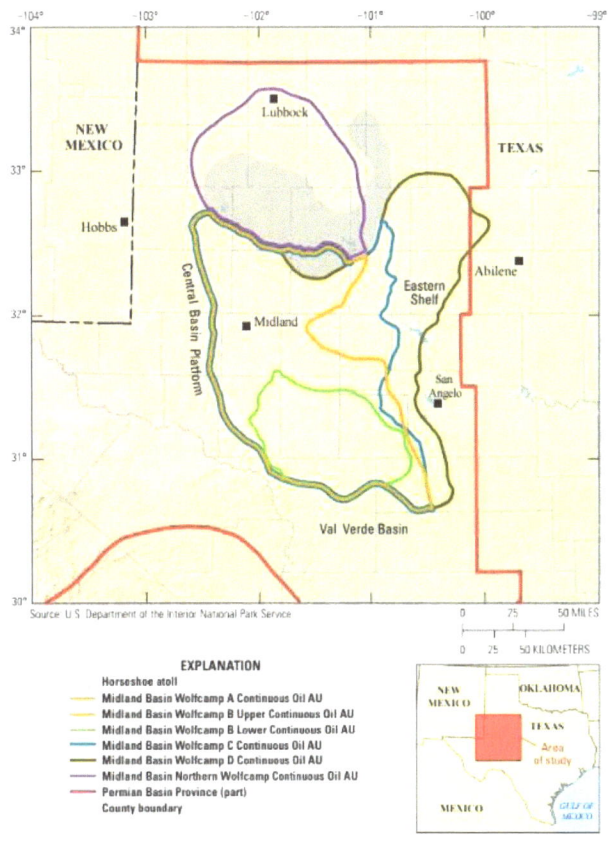

encompasses Mississippian, Pennsylvanian, and Lower Permian reservoirs. Oil has been produced using traditional vertical well technology.

However, more recently, oil and gas companies have been using horizontal drilling and hydraulic fracturing, and more than 3,000 horizontal wells have been drilled and completed in the Midland Basin Wolfcamp section.

The Wolfcamp shale is also present in the Delaware Basin portion of the Permian Basin province, but was not included in this assessment. The Permian Basin province includes a series of basins and other geologic formations in West Texas and southern New Mexico. It is one of the most productive areas for oil and gas in the entire United States.

Continuous oil and gas is dispersed throughout a geologic formation rather than existing as discrete, localized occurrences, such as those

December 2016 • Issue 12

Table 1. Key assessment input data for six continuous assessment units in the Wolfcamp shale in the Midland Basin of the Permian Basin Province, Texas.

AU, assessment unit; %, percent; EUR, estimated ultimate recovery per well; MMBO, million barrels of oil. The average EUR input is the minimum, median, and calculated mean. Shading indicates not applicable.

Assessment input data	Midland Basin Wolfcamp A Continuous Oil AU				Midland Basin Wolfcamp B Upper Continuous Oil AU			
	Minimum	Mode	Maximum	Calculated mean	Minimum	Mode	Maximum	Calculated mean
Potential production area of AU (acres)	3,000,000	3,495,500	5,814,000	4,103,000	3,000,000	3,523,000	5,814,000	4,112,333
Average drainage area of wells (acres)	60	80	160	100	60	80	160	100
Percentage of area untested in AU	83	85	91	86.3	83	85	91	86.3
Success ratios (%)	92	95	99	95.3	92	95	99	95.3
Average EUR (MMBO)	0.12	0.16	0.3	0.167	0.12	0.16	0.3	0.167
AU probability	1.0				1.0			
Assessment input data	**Midland Basin Wolfcamp B Lower**				**Midland Basin Wolfcamp C Continuous**			
	Minimum	Mode	Maximum	Calculated mean	Minimum	Mode	Maximum	Calculated mean
Potential production area of AU (acres)	700,000	742,000	2,012,000	1,151,333	1,000,000	2,373,000	6,703,000	3,358,667
Average drainage area of wells (acres)	60	80	160	100	60	80	160	100
Percentage of area untested in AU	68	70	89	75.7	47	77	92	72
Success ratios (%)	92	95	99	95.3	50	70	90	70
Average EUR (MMBO)	0.12	0.16	0.3	0.167	0.03	0.08	0.15	0.083
AU probability	1.0				1.0			
Assessment input data	**Midland Basin Wolfcamp D Continuous**				**Midland Basin Northern Wolfcamp**			
	Minimum	Mode	Maximum	Calculated mean	Minimum	Mode	Maximum	Calculated mean
Potential production area of AU (acres)	2,000,000	4,885,000	8,915,000	5,266,667	1,000	1,633,000	3,266,000	1,633,333
Average drainage area of wells (acres)	60	80	160	100	60	80	160	100
Percentage of area untested in AU	73	89	94	85.3	97	99	100	98.7
Success ratios (%)	75	85	95	85	10	50	90	50
Average EUR (MMBO)	0.06	0.12	0.25	0.126	0.02	0.06	0.14	0.064
AU probability	1.0				1.0			

in conventional accumulations. Because of that, continuous resources commonly require special technical drilling and recovery methods, such as hydraulic fracturing.

Undiscovered resources are those that are estimated to exist based on geologic knowledge and theory, while technically recoverable resources are those that can be produced using currently available technology and industry practices. Whether or not it is profitable to produce these resources has not been evaluated.

USGS is the only provider of publicly available estimates of undiscovered technically recoverable oil and gas resources of onshore lands and offshore state waters. The USGS Wolfcamp shale assessment was undertaken as part of a nationwide project assessing domestic petroleum basins using standardized methodology and protocol.

Geologic Summary

The Permian Basin Province of west Texas and southeastern New Mexico contains two sub basins, the Delaware Basin to the west and the Midland Basin to the east, separated by the uplifted

December 2016 • Issue 12

Table 2. Assessment results for six continuous assessment units in the Wolfcamp shale in the Midland Basin of the Permian Basin Province, Texas.

MMBO, million barrels of oil; BCFG, billion cubic feet of gas; MMBNGL, million barrels of natural gas liquids. Results shown are fully risked estimates. For gas accumulations, all liquids are included under the natural gas liquids NGL category. F95 represents a 95-percent chance of at least the amount tabulated. Other fractiles are defined similarly. Fractiles are additive under the assumption of perfect positive correlation. Shading indicates not applicable

Total petroleum system and assessment unit (AU)	AU probability	Accumulation type	Total											
			Oil				Gas				NGL (MMBNGL)			
			F95	F50	F5	Mean	F95	F50	F5	Mean	F95	F50	F5	Mean
Permian Basin Paleozoic Composite Total Petroleum System														
Midland Basin Wolfcamp A Continuous Oil AU	1.0	Oil	3,754	5,633	8,483	5,815	2,540	4,453	7,457	4,652	223	436	806	465
Midland Basin Wolfcamp B Upper Continuous Oil AU	1.0	Oil	3,769	5,644	8,505	5,829	2,557	4,454	7,482	4,663	224	437	811	466
Midland Basin Wolfcamp B Lower Continuous Oil AU	1.0	Oil	794	1,342	2,351	1,430	554	1,056	2,023	1,144	49	104	215	114
Midland Basin Wolfcamp C Continuous Oil AU	1.0	Oil	577	1,306	2,728	1,433	417	1,018	2,299	1,146	38	100	241	115
Midland Basin Wolfcamp D Continuous Oil AU	1.0	Oil	2,420	4,658	8,262	4,920	1,733	3,657	7,096	3,936	156	357	753	394
Midland Basin Northern Wolfcamp Continuous Oil AU	1.0	Oil	116	458	1,139	521	86	357	953	417	8	35	100	42
Total undiscovered continuous resources			**11,430**	**19,041**	**31,468**	**19,948**	**7,887**	**14,995**	**27,310**	**15,958**	**698**	**1,469**	**2,926**	**1,596**

Central Basin platform. During the Pennsylvanian-Permian, the Wolfcamp was deposited as shallow-water carbonates on the Central Basin platform and Eastern shelf and interbedded, finer-grained, organic-rich siliciclastic mud with organic-poor, clay- rich mud and fine-grained carbonates in the deeper part of the Midland Basin.

The Wolfcamp shale throughout this region is mature for oil generation based on thermal maturation data, Pawlewicz and others, 2005. The petroleum industry has divided the Wolfcamp shale into four stratigraphic units based on petrophysical log signatures and landing zones for horizontal wells.

The uppermost unit is the Wolfcamp A, followed by the underlying Wolfcamp B, C, and D units, respectively. The eastern margin of Wolfcamp shale deposition pro- graded westward through time, as indicated by the larger depositional areas delineated in the Wolfcamp C and D assessment units AUs when compared to the Wolfcamp A and B assessment units.

Definition of Assessment Units

Six continuous assessment units were defined and quantitatively assessed in the Wolfcamp shale in the Midland Basin of the Permian Basin Province:

- 1 Midland Basin Wolfcamp A Continuous Oil AU,
- 2 Midland Basin Wolfcamp B Upper Continuous Oil AU,
- 3 Midland Basin Wolfcamp B Lower Continuous Oil AU,
- 4 Midland Basin Wolfcamp C Continuous Oil AU,
- 5 Midland Basin Wolfcamp D Continuous Oil AU, and
- 6 Midland Basin Northern Wolfcamp Continuous Oil AU. see map

The six assessment units are within the Permian Basin Paleozoic Composite Total Petroleum System TPS, Schenk and others, 2007.

The Midland Basin Wolfcamp A Continuous Oil AU

Continued on page 19

P O G S March 2017

The Intl Pipeline Oil and GAS Safety Conference and Exhibition

March 14-16, 2017 Houston Texas USA

Organized by:

Pipeline Integrity | **Emission Reduction** | **Well Control** | **Oil and Gas Transportation** | **Chemical Extraction**

Connecting Supplier with Procurement Teams

Exhibition

200+

Exhibitors Expected

Attendance

2000+

Attendees Expected

Goal

Improve safety in the entire value chain of the oil and gas industry not limited to the well heads but distribution chains, transportation and supply chain.

Exhibit@ P O G S Safety Tech

P O G S Safety Tech provides international and local energy companies who operate across the up, mid and downstream sectors of the oil &gas supply chain with a B2B platform to meet and influence highly-focused International decision-makers and buyers.

Who is Attending?

Take Advantage of Early Registration- Register Now @

http://www.oilandgassafetyconference.com
registration/online-registration/

Who is Exhibiting?

SHOWFLOOR IS Selling Very Fast
RESERVED Today

http://www.oilandgassafetyconference.com/
booth-registeration/

Official Media Partner

For further details visit website @
http://oilandgassafetyconference.com
or call +1-832-664-0618

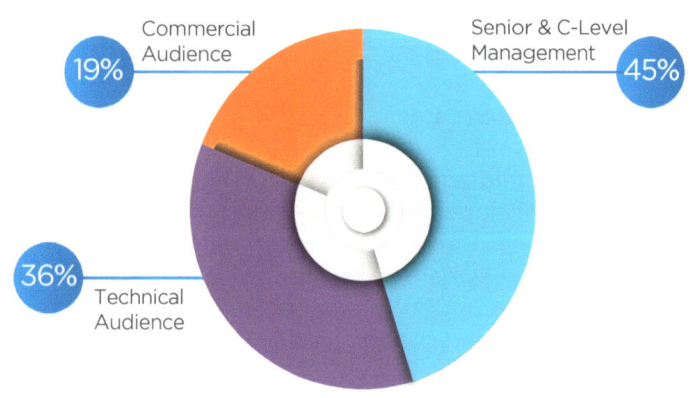

Commercial Audience 19%

Senior & C-Level Management 45%

Technical Audience 36%

The International Pipeline,
Oil and Gas Safety Conference

Intl Pipeline, Oil and Gas Safety Conference & Exhibition

POGS March 14-16, 2017   www.oilandgassafetyconference.com

Houston Astros - Minute Maid Park - 501 Crawford Street | Houston, Texas 77002

Early Discount Ends December 28, 2016

Goals

This conference seeks to address process safety issues in the upstream, midstream and downstream subsectors of the industry; with special focus on well control safety, process safety, pipeline safety, and new regulatory impact.

To help improve operational excellence in the various communities where the industry operates- emerging technologies, leak detection and prevention technologies, emission reduction technologies, compliance audit, best practices to reduce risks and hazards, and improve the overall operational safety is the focal point of this conference.

To help meet these goals - are the speakers and participating companies

Brady Austin
QHSE Service Line Owner Lloyd's Register

Mothusi Pahl
Vice President- Alphabet Energy Inc.

Vincent Higgins
Chairman and CEO Optech4D Inc

Hunter Hawa
Global EHS Director for PSRG

Robert Miller
Regulatory Compliance Specialist, Veriforce

W. Duncan Welder IV
RISC's Director of Client Services

Shoshi Kaganovsky
CEO and founder of SensoLeak

Alexis Vitone
President, AvA Excellence in Business Strategies & HSE, LLC

Tom Meek
Director of Compliance, Veriforce

Keith J. Coyle
Shareholder, Babst Calland

Mark A. Hernandez
President of Multiply Leadership

Rixio Medina
Director of Business Development for the Board of Certified Safety Professionals

Early Registration Fee - $350

Register Today for this all important industrial conference

Fill out this form email form to: *registration@oilandgassafetyconference.com*

Or mail form with check to the address below.

Mail and make check payable to: *RGT Media Communications Corp.   10777 Westheimer Street, #1100 Houston Texas 77042*

Payment Method -Card type- Amex, Visa, Master, Discovery (circle one)

Card No: _____   Expiration Date: _____   Name on card: _____   By Check Check No: _____

First Name: _____   Last Name: _____

Your Preferred Mailing Address - (Circle One) Business/ Residence

Job Title: _____   Company Name : _____   Street : _____

(No PO Boxes Please) City : _____   State: _____   Country: _____   Zip/Postal Code: _____

Day Phone: _____   Fax: _____   E-mail: _____

cut here

Program Agenda Break Down

Pipeline Safety- Leak Detection and Prevention Tech

Shoshi Kaganovsky - CEO and founder of SensoLeak

Emerging Technologies - Leveraging Virtual and Augmented Reality Technologies for Midstream & pipeline industries

Vincent Higgins - Chairman and CEO Optech4D Inc

Best Practices- Avoiding risks and hazards/ Competency-Based Training Program

Alexis Vitone- President - AvA Excellence in Business Strategies & HSE, LLC

Brady Austin - QHSE Service Line Owner- Lloyd's Register

W. Duncan Welder IV - RISC's Director of Client Services

Motivational Speaker

Mark A. Hernandez - President Multiply Leardership

Process Safety

Hunter Hawa - Global EHS Director for PSRG

PHMSA Regulations

Keith J. Coyle - Shareholder- Attorney at Law - Babst Calland

Emission Reduction Technology- Converting Flares to Power Gen

Mothusi Pahl - Vice President-Alphabet Energy Inc.

Compliance Audit- Federal/State codes and OQ NPRM

Tom Meek - Director of Compliance, Veriforce

Robert Miller - Regulatory Compliance Specialist, Veriforce

Rixio Medina - Director of Business Development for the Board of Certified Safety Professionals

---

Supporting Organization
Pennsylvania Independent Oil and Gas Association
PIOGA

Official Media Partner
USA Oil and Gas Monitor

Member Organization
Independent Petroleum Association of America
IPAA

POGS
Int'l Pipeline, Oil and Gas
Safety Conference &
Exhibition

cut
here

Contined from page 15

and Wolfcamp B Upper Continuous Oil AU were constrained by the Central Basin platform to the west, the southern edge of the Horseshoe atoll and an area of thick Wolfcamp strata to the north, the Eastern Shelf margin to the east, and the Val Verde Basin Canyon Sandstone AU, Schenk and others, 2016, to the south.

The Midland Basin Wolfcamp B Lower Continuous Oil AU is restricted to the southern Midland Basin where the thickness of the Wolfcamp B exceeds 500 feet and allows for placement of two lateral wells in the Wolfcamp B unit.

The Midland Basin Wolfcamp C Continuous Oil AU is bounded by the Central Basin platform on the west, the southern edge of the Horseshoe atoll and an area of thick Wolfcamp strata to the north, the estimated extent of the Wolfcamp C depositional boundary using shelf edges mapped by Hentz and

others 2016, to the east, and the Val Verde Basin Canyon Sandstone AU, Schenk and others, 2016, to the south.

The Midland Basin Wolfcamp D Continuous Oil AU is defined by the Central Basin platform to the west, the southern edge of the Horseshoe atoll to the north, the mapped extent of Upper Pennsylvanian shale to the northeast and east, and the Val Verde Basin Canyon Sandstone AU, Schenk and others, 2016, to the south.

The Midland Basin Northern Wolfcamp AU is defined by the Midland Basin shelf margin to the north and east and the southern edge of the Horseshoe atoll and the northern boundary of an area of thick Wolfcamp strata that was included in the AUs to the south. Assessment input data for the six AUs are summarized in table 1.

USA Oil and Gas Monitor

A RGT Media Communications Corp.

SUBSCRIPTION FORM

First Name _____   Middle _____   Last _____

Current Job Title _____   Job Title Code _____

Company Name _____

Preferred Mailing Address - (Circle One)

    Business   Residence

Street _____ (No PO Boxes Please)

City _____   State _____   Zip _____

Country _____

Day Phone _____   If outside U.S., include country code. (ex: 000-000-000-0000)

Fax _____   Email _____

Form Instructions:

Email completed form to subscribe@usaoilandgasmonitor.com or mail form with check to the address below.

RGT Media Communications Corp.
10777 Westheimer Road #1100
Houston Texas 77042

1 Year Digital Subscription

For non-Texas subscribers - $119.88

Subscribers living in Texas – pays $119.88 plus 8.25% state tax $9.89 = $129.77

1 Year Print Subscription

For non-Texas subscribers - $144

Please add shipping cost and multiply by 12 (for example $1.67 x12) = 20.04

Subscribers living in Texas – pays $144 plus 8.25% state tax $11.88= $155.88

Please add shipping cost and multiply by 12 (for example $1.67 x12) = 20.04

Shipping Cost (calculated by weight)

Circle choice from the following option and add to the subscription cost

First Class 1- 5 business/days = $1.67

Fedex Shipping 1-3 business/days= $6.40

USPS Priority 1-3 business/days= $3.56

International First Class 1-7 business/days= $12.44

You can also pay for subscription online by visiting our website:
www.usaoilandgasmonitor.com/subscribe
Wire transfer, call Jewel Spring, 832-486-0095 for any questions.

Payment Method

Card Type (circle one)

Amex   Visa   Master   Discovery

Card No.

_____

Expiration Date

_____

CSV No.

_____

Name on Card

_____

By Check

Check No.

_____

December 2016 • Issue 12

Baker Hughes now a GE Company
World Leader in Oil and Gas Productivity

GE and Baker Hughes has announced that the companies have entered into an agreement to combine GE's oil and gas business and Baker Hughes to create a world-leading oilfield technology provider with a unique mix of service and equipment capabilities. The "New" Baker Hughes will be a leading equipment, technology and services provider in the oil and gas industry with $32 billion of combined revenue and operations in more than 120 countries. By drawing from GE technology expertise and Baker Hughes capabilities in oilfield services, the new company will provide best-in-class physical and digital technology solutions for customer productivity.

Under the terms of the agreement, which has been unanimously approved by the boards of directors of both companies, at the closing of the transaction Baker Hughes shareholders will receive a special one-time cash dividend of $17.50 per share and 37.5 per cent of the new company. GE will own 62.5 per cent of the company. The transaction is expected to close in mid-2017.

"This transaction creates an industry leader, one that is ideally positioned to grow in any market. Oil & gas customers demand more productive solutions. This can only be achieved through technical innovation and service execution, the hallmarks of GE and Baker Hughes," said Jeff Immelt, Chairman and Chief Executive Officer of GE. "As we built the GE Oil & Gas business, I have always been impressed by the respect our customers have for Baker Hughes. GE Oil & Gas is a key GE business, one that fully leverages the GE Store. As we go forward, this transaction accelerates our capability to extend the digital framework to the oil and gas industry. An oilfield service platform is essential to deliver digitally enabled offerings to our customers. We expect Predix to become an industry standard and synonymous with improved customer outcomes.

- Highly complementary transaction combines GE's oil and gas technology, manufacturing and digital platform with Baker Hughes' oilfield services offerings and technologies
- Combination creates an unparalleled company positioned to deliver value for customers and investors
- GE to own 62.5per cent and Baker Hughes shareholders to own 37.5per cent of the "New" Baker Hughes
- GE to contribute $7.4 billion to fund the $17.50 per share special dividend to existing Baker Hughes shareholders
- Expected to be accretive to GE 2018 earnings per share by approximately $.04; Synergies of $1.6 billion expected to be realized by 2020
- Lorenzo Simonelli will be CEO, Jeff Immelt will be Chairman and Martin Craighead will be Vice Chairman of the "New" Baker Hughes Board of Directors

GE investors will benefit through ownership of a stronger business with substantial synergies and an improved competitive position. The transaction is expected to add approximately $.04 to GE EPS in 2018, $.08 by 2020."

Martin Craighead, Chairman and Chief Executive Officer at Baker Hughes said, "This compelling combination brings together best-in-class oilfield equipment manufacturing and services, and digital technology offerings for the benefit of all customers and stakeholders. The combination of our complementary assets will create a platform capable of seamless integration while we enhance our ability to deliver optimized and integrated solutions and increase touch points with our customers. In addition, Baker Hughes shareholders will receive a special one-time cash dividend of $17.50 per share and benefit from the upside of a stronger, larger business. With employees of Baker Hughes and GE Oil & Gas coming together, the new

December 2016 • Issue 12

company will be an industry leader, well-positioned to compete in the oil and gas industry while pushing the boundaries of innovation for our customers."

Lorenzo Simonelli, who is currently president and CEO of GE Oil & Gas said, "This transformative transaction will create a powerful force in the oil and gas market as we continue to drive long-term value for our customers and shareholders. This transaction is also exciting for employees of both companies. GE Oil & Gas and Baker Hughes are an exceptional cultural fit, sharing a commitment to exceeding customer expectations. Both companies' employees will benefit significantly from being part of a larger, stronger company that is positioned for long-term growth. We look forward to combining the digital solutions and technology from the GE Store with the domain expertise of Baker Hughes and its culture of innovation in the oilfield services sector."

Compelling Strategic and Financial Benefits of the Transaction

Complementary assets and integrated offerings will provide differentiated services for combined company's customers. The company will combine the digital solutions, manufacturing expertise and technology from the GE Store and the outstanding track record of success Baker Hughes has in the oilfield services sector. With combined revenue of over $32 billion1 the product portfolio of GE Oil & Gas and Baker Hughes in drilling, completions, production and midstream / downstream equipment and services will create the second largest player in the oilfield equipment and services industry. Customers should expect sustainable innovation and integration that will deliver valuable outcomes. As one company, we will have operations in more than 120 countries. Both companies have invested even in the downturn and have strong, complementary competitive scope across the industry. From GE's full stream oil and gas manufacturing and technology solutions spanning across subsea & drilling, rotating equipment, imaging and sensing to the Baker Hughes portfolio in Drilling & Evaluation and Completion & Production, the combined company will be moving beyond oilfield services and into oil and gas productivity solutions.

- **The combination produces substantial synergies through combined efficiency and growth.** The companies expect to generate total run rate synergies of $1.6 billion by 2020, which has a net present value of $14 billion. While this is primarily driven by cost out, we believe that the new company is positioned for growth as the industry rebounds.

- **Combination positioned to create value for Baker Hughes shareholders.** The diversified portfolio can deliver through the oil and gas cycle. There is a large pool of synergies that will improve operating margins and drive organic growth. The "New" Baker Hughes has a strong balance sheet.
- **Combination positioned to create value for GE shareholders.** The transaction is expected to be accretive to GE's earnings per share by $.04 by 2018 and $.08 by 2020. This is another step in creating the premium digital industrial company.
- **The "New" Baker Hughes is expected to be the partner and employer of choice for the industry.** Combination is an exceptional cultural fit. Both companies' employees will benefit significantly from being part of a larger, more diversified company.

Financial Structure
The transaction will be executed using a partnership structure, pursuant to which GE Oil & Gas and Baker Hughes will each contribute their operating assets to a newly formed partnership. GE will have a 62.5 per cent interest in this partnership and existing Baker Hughes shareholders will have a 37.5 per cent interest through a newly NYSE listed corporation. Baker Hughes shareholders will also receive a special one-time cash dividend of $17.50 per share at closing. The $7.4 billion contributed by GE to the new partnership will be used to fund the cash dividend to existing Baker Hughes shareholders.

Headquarters, Management and Board of Directors
The "New" Baker Hughes will have dual headquarters in Houston, Texas and London, UK.

Jeff Immelt, Chairman and CEO of GE will serve as Chairman of the Board of Directors and Lorenzo Simonelli, president and CEO of GE Oil & Gas will serve as President and Chief Executive Officer. Martin Craighead, Baker Hughes Chairman and CEO, will serve as Vice Chairman of the Board. The remainder of the executive leadership team will be a combination of existing leaders from both GE and Baker Hughes.

Baker Hughes is a leading supplier of oilfield services, products, technology and systems to the worldwide oil and natural gas industry. The company's 34,000 employees today work in more than 80 countries helping customers find, evaluate, drill, produce, transport and process hydrocarbon resources.

Cybersecurity Due Diligence Is Crucial in All M and A—Including Energy M and A Transactions

A Cautionary Tale in Cybersecurity Due Diligence

Can a single data breach kill or sideline a deal? Perhaps so. Last month Verizon signaled that Yahoo!'s disclosure of a 2014 cyberattack might be a "material" change to its July $4.83 billion takeover bid—which could lead Verizon to renegotiate or even drop the deal entirely. Concern over cybersecurity issues is not unique to technology or telecommunications combinations. In a 2016 NYSE Governance Services survey of public company directors and officers, only 26 per cent of respondents would consider acquiring a company that recently suffered a high-profile data breach—while 85 per cent of respondents claimed that it was "very" or "somewhat" likely that a major security vulnerability would affect a merger or acquisition under their watch e.g., 52 per cent said it would significantly lower valuation.

Bottom Line: Cybersecurity should play a more meaningful role in the due diligence portion of any potential M and A deal. Certainly, this is so when a material portion of the value in the acquisition comes

from intangible assets that might be most vulnerable to hackers. Financial information comes to mind. Personal information of employees does as well. But companies also need to be concerned about their trade secrets, know-how and other confidential business information whose value inheres in its secrecy. Therefore, a merely perfunctory approach to cybersecurity can become very costly. The union of companies today is a union of information, malware and all.

Energy M and A Is Not Immune

To weather the plunge in prices, many oil companies have sought out new innovations to reduce the cost of extraction and exploration. Investments in digital technologies will likely only increase—a 2015 Microsoft and Accenture survey of oil and gas industry professionals found that "Big Data" and the "Industrial Internet of Things" IIoT are targets for greater spend in the next three to five years. Cybersecurity threats were perceived in the survey as one of the top two barriers to realizing value from these technologies.

These developments in energy industry—bigger data and bigger vulnerabilities—are here to stay. The

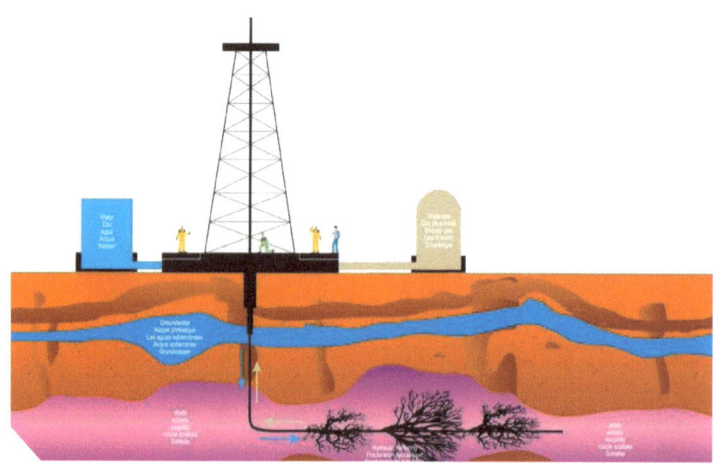

merger of General Electric and Baker Hughes also speaks to the growing importance of analytics to oil production. Commentators note that the acquisition would allow GE more fully to implement its Predix platform, an application of IIoT to connect everything from wellhead sensors to spreadsheets. However, as last month's massive cyberattack on DNS provider Dyn, Inc. demonstrated, the IIoT holds unique challenges as well as great promise for operational efficiency. In this attack, reportedly 400,000 internet-linked gadgets were hacked and used to reroute web traffic to overload servers.

Bottom Line: Robust cybersecurity diligence should be de rigueur for energy M&A.

What Can Companies Do to Protect Deal Value?

For starters, energy companies should treat cybersecurity as a separate and more involved category for due diligence.

Liability for or damages from legacy data breaches or malware can become expensive—damages to systems, theft of information and liability from the release of personal or reputation-damaging information, to name a few. Therefore, anticipating problems post-merger, cataloguing past vulnerabilities and most importantly, discovering actual breaches before closing is crucial to avoid deals blowing hot and cold.

Companies should retain IT specialists who can do an objective assessment of the cybersecurity posture of a proposed merger or acquisition. This can help prospective acquirers better determine the adequacy of a target's cybersecurity programs, such as its policies over incident response, how access to data is distributed, the extent of a company's online presence and vulnerabilities, and how remediation of any potential cyberthreats or actual breaches may best proceed.

A cybersecurity questionnaire should also be developed, covering such topics as:

How and where has company data been stored?

Who has had access?

Have there been any actual or attempted intrusions into or leaks of company data?

An acquirer could further insist on specific representations and warranties from a target company regarding their cybersecurity compliance, as well as bargain towards indemnity for prior data breaches.

On the target side, energy companies should prepare in turn for more scrutiny over their data security and privacy practices. Among other benefits to "knowing thyself," getting ahead of this process should offer targeted companies a better negotiating position. It would also allow them to take a more proactive role in defining the policies of the combined company post-merger. At the very least, these efforts could help avoid the kind of hiccups and uncertainties that lead to undervaluation. In any event, poor cybersecurity practices can give an impression that a target lacks risk management in other areas—not an ideal pose to strike in any bargain.

Parting Thoughts

It is a trope in cybersecurity writing to invoke figures like Sun Tzu and shoehorn in quotes about war stratagem. Well, these habits are in some ways unavoidable: For all intents and purposes, fighting anonymous hackers resembles battle prep—a method of self-awareness and readiness that defies box-checking.

Energy companies could take these words to heart from the inestimable Miyamoto Musashi, a samurai who won 60 duels: "If you consciously try to thwart opponents, you are already late." A sentiment echoed more recently by Mike Tyson's truistic "Everyone has a plan until they get punched in the mouth."

And This Key Takeaway: Any cybersecurity program must go hand-in-hand with a corporate culture that respects data as among its most valued assets. Efforts in detection, reporting and remediation are challenges that fall throughout the ranks and, if reflexive to the unknown, stand the best chance of being fully realized.

Bottom Line: Mind Your Data!

By: Jeff C. Dodd, Doug Rommelmann and Ross Campbell

December 2013 • Issue 12

Dr. Fatih Birol sees Natural Gas, Wind and Solar Replacing Coal in The Next 25 Years

As a result of major transformations in the global energy system that take place over the next decades, renewables and natural gas are the big winners in the race to meet energy demand growth until 2040, according to the latest edition of the World Energy Outlook, the International Energy Agency's flagship publication.

Dr Fatih Birol, the IEA's executive director said, "We see clear winners for the next 25 years - natural gas but especially wind and solar - replacing the champion of the previous 25 years, coal. But there is no single story about the future of global energy: in practice, government policies will determine where we go from here."

This transformation of the global energy mix described in WEO-2016 means that risks to energy security also evolve. Traditional concerns related to oil and gas supply remain – and are reinforced by record falls in investment levels. The report shows that another year of lower upstream oil investment in 2017 would create a significant risk of a shortfall in new conventional supply within a few years.

In the longer-term, investment in oil and gas remain essential to meet demand and replace declining production, but the growth in renewables and energy efficiency lessens the call on oil and gas imports in many countries. Increased LNG shipments also change how gas security is perceived. At the same time, the variable nature of renewables in power generation, especially wind and solar, entails a new focus on electricity security.

"We are entering a period of greater oil price volatility," said Dr Birol. "If oil prices rise in the short term, then shale producers can react quite quickly to put more oil on the market, producing a see-saw movement. And if we continue to see subdued investments in new conventional oil projects, this could have profound consequences in the longer term."

Global oil demand continues to grow until 2040, mostly because of the lack of easy alternatives to oil in road freight, aviation and petrochemicals, according to WEO-2016. However, oil demand from passenger cars declines even as the number of vehicles doubles in the next quarter century, thanks mainly to improvements in efficiency, but also biofuels and rising ownership of electric cars.

Coal consumption barely grows in the next 25 years, as demand in China starts to fall back thanks to efforts to fight air pollution and diversify the fuel mix. The gas market is also changing, with the share of LNG overtaking pipelines and growing to more than half of the global long-distance gas trade, up from a quarter in 2000. In an already well-supplied market, new LNG from Australia, the United States and elsewhere triggers a shift to more competitive markets and changes in contractual terms and pricing.

The Paris Agreement, which entered into force on 4 November, is a major step forward in the fight against global warming. But meeting more ambitious climate goals will be extremely challenging and require a step change in the pace of decarbonization and efficiency. Implementing current international pledges will only slow down the projected rise in energy-related carbon emissions from an average of 650 million tons per year since 2000 to around 150 million tons per year in 2040.

While this is a significant achievement, it is far from enough to avoid the worst impact of climate change as it would only limit the rise in average global temperatures to 2.7°C by 2100. The path to 2°C is tough, but it can be achieved if policies to accelerate further low carbon technologies and energy efficiency are put in place across all sectors.

It would require that carbon emissions peak in the next few years and that the global economy becomes carbon neutral by the end of the century. For example, in the WEO-2016 2°C scenario, the number of electric cars would need to exceed 700 million by 2040, and displace more than 6 million barrels a day of oil demand. Ambitions to further limit temperature gains, beyond 2°C, would require even bigger efforts.

"Renewables make very large strides in coming decades but their gains remain largely confined to electricity generation," said Dr Birol. "The next frontier for the renewable story is to expand their use in the industrial, building and transportation sectors where enormous potential for growth exists."

December 2016 • Issue 12

New Pipeline Infrastructure Projects to Increase Natural Gas Takeaway Capacity In 2017

A number of pipeline projects that have been approved, or are in various stages of the approval process, would increase capacity to transport natural gas from the Utica production region in Ohio to natural gas markets. Collectively, these projects could add up to 6.8 billion cubic feet per day Bcf/d of takeaway capacity out of the Utica region by the end of 2018.

Over the past several years, natural gas production in the Appalachian basin from the Marcellus and Utica shales has grown significantly. Because pipeline projects often have longer lead times than production projects, transport infrastructure for accessing natural gas demand centers and export locations in the Appalachian Basin has not kept pace with production capability. This situation has resulted in a lower price for natural gas from the Appalachian region relative to many other natural gas trading hubs in the United States.

Construction of a new interstate natural gas pipeline in the United States requires approval by the Federal Energy Regulatory Commission FERC, which can be a lengthy process. Construction on a pipeline can begin once a final environmental impact statement is issued, pending that project receiving Clean Water Act, Coastal Zone Management Act, Clean Air Act, and other necessary state permits.

Map 1- Federal Energy Regulatory Commission FERC Process for Natural Gas Certificates

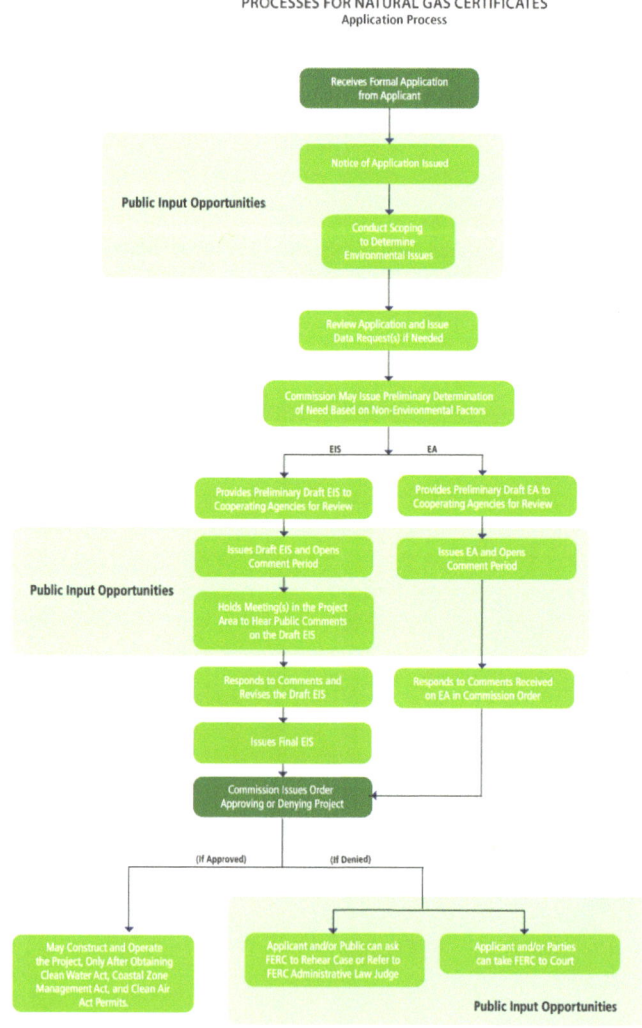

PROCESSES FOR NATURAL GAS CERTIFICATES
Application Process

December 2016 • Issue 12

Pipeline Operator Name	Project Type	Status	Year In Service Date	State(s)
Kaiser-Frontier Midstream	New Pipeline	Approved	na	WY, CO
Sawgrass Storage LLC	Lateral	Approved	na	LA
DCP Midstream	Lateral	Approved	na	CO
Dominion Cove Point LNG	Expansion	Approved	na	MD
Dominion Cove Point LNG	Expansion	Approved	na	MD
Constitution Pipeline Co	New Pipeline	Approved	2017	PA, NY
McMoran Exploration Inc	New Pipeline	Approved	2020	GM, AL
Port Dolphin Pipeline LP	New Pipeline	Approved	2018	FL
Transcontinental Gas Pipeline	Reversal	Approved	2017	PA, VA, NC, SC, GA, AL
Transcontinental Gas Pipeline	Expansion	Approved	2017	LA
Transcontinental Gas Pipeline	Expansion	Approved	2017	AL, GA
Transcontinental Gas Pipeline	Expansion	Approved	2017	AL
Algonquin Gas Transmission	Lateral	Approved	2017	MA
Texas Eastern Transmission co	Reversal	Approved	2017	PA, OH, WV, KY, TX
ET Rover Pipeline	New Pipeline	Approved	2017	PA, WV, OH, MI, CN
Tennessee Gas Pipeline	Expansion	Approved	2017	WV
Columbia Gulf Transmission	Reversal	Approved	2017	LA
Transcontinental Gas Pipeline	Expansion	Approved	2017	LA
Transcontinental Gas Pipeline	Expansion	Approved	2017	NJ
NextEra Energy	New Pipeline	Approved	2017	FL
Spectra Energy Corp/NextEra Energy, Inc	New Pipeline	Approved	2017	AL, GA, FL
Panhandle Eastern Pipeline	Reversal	Approved	2017	OH, IN, IL
Tennessee Gas Pipeline	Expansion	Approved	2016	NY
Transcontinental Gas Pipeline	Expansion	Approved	2016	NJ
East Tennessee Natural Gas	Expansion	Approved	2016	TN
Dominion Carolina Gas Transmission	New Pipeline	Approved	2016	SC
Cheniere Creole Trail Pipeline	Reversal	Approved	2016	LA
Impulsora Pipeline	New Pipeline	Approved	2016	TX, MX
Florida Gas Transmission Company	Expansion	Approved	2016	FL, NY
Transwestern Pipeline	Expansion	Approved	2016	NM
Dominion Transmission	New Pipeline	Approved	2016	WV
MoBay Storage Hub	Lateral	Approved	2016	AL
Dominion Transmission	Expansion	Approved	2016	PA, OH

Key projects that are undergoing FERC review and may enter service in the next few years include:

The **Rover pipeline**, which recently received a final environmental impact statement from FERC, is designed to transport 3.25 Bcf/d of natural gas from the Marcellus and Utica Shale areas to various market hubs.

The **Leach Xpress project**, which received a draft environmental impact statement DEIS from FERC, seeks to add 1.5 Bcf/d of natural gas takeaway capacity along the Columbia Pipeline Group's network.

The **Rayne Xpress project**, which received a DEIS,

will augment the Leach Xpress project. The Rayne Xpress Project seeks to add 0.6 Bcf/d in takeaway capacity from the Columbia Pipeline system to Gulf Coast markets, which will help facilitate liquefied natural gas exports, among other uses.

The **Nexus Gas Transmission project**, which received a DEIS from FERC in July 2016, is designed to deliver 1.5 Bcf/d of natural gas supplies from the Utica region to markets in northern Ohio, southeastern Michigan, the Chicago Hub in Illinois, and the Dawn Hub in Ontario, Canada.

Other Key Projects that will come onstream in 2017

Constitution Pipeline Co, Florida Southeast Connection, and Sabal Trail Project are new pipeline

December 2016 • Issue 12

projects that have been approved and will come on stream in 2017. See Table:

Project Type	Definition
Conversion	Pipelines that were converted from transporting other sources to natural gas
Expansion	Projects that expanded mainline capacity or mileage including additional compressors, looping, or extensions.
Lateral	Projects that add lateral lines connecting the mainline to power plants, processing plants, industrial plants, storage facilities and other pipelines.
New Pipeline	New pipeline system
Upgrade	Projects that upgrade the pipelines including upgrading compressions or pipelines

The First Microchannel Gas-To-Liquid Plants
Convert Stranded Natural Gas to Marketable Products

The first microchannel gas-to-liquid GTL plant in the United States was completed in September. The new plant, built by ENVIA Energy, is located in Oklahoma and is expected to begin converting landfill gas into liquid petroleum products later this year.

GTL plants convert natural gas to higher-valued petroleum products, including liquid fuels, waxes, and chemical feedstocks. The most common conversion method is the Fischer-Tropsch F-T process, which involves a series of chemical reactions that transform natural gas, or a gasified solid fuel, such as coal or biomass, into hydrocarbons and water. Six large-scale F-T GTL plants operate in the world today: two in South Africa, two in Qatar, and one each in Malaysia and Nigeria. These plants have output capacities ranging from 5,600 barrels per day b/d to 140,000 b/d. BP operated a smaller, 300 b/d pilot plant in Alaska from 2002 to 2009, but no commercial-scale GTL plants currently operate in the United States.

Once commissioned, ENVIA Energy's plant will have a capacity of 300 b/d. For comparison, a petroleum refinery on the U.S. Gulf Coast may have a capacity of tens or hundreds of thousands of barrels per day. The project is a joint venture between four companies that plan to build more microchannel GTL plants at landfill sites. Several other companies are also developing microchannel GTL plants in the United States, including a 100 b/d plant scheduled to be completed next month in Wharton, Texas.

Because F-T reactions require high temperature and pressure, building a suitable reaction vessel can be expensive. High capital costs, coupled with market uncertainty regarding natural gas and petroleum product prices, has led several companies to develop different techniques.

ENVIA Energy's microchannel gas-to-liquids plant

High temperatures and pressures are less costly to maintain at smaller volumes. Small-scale F-T GTL plants can use microchannel reactors, diameters of one millimeter or less, to optimize their operation. The small diameters of the reactor vessels allow for better temperature control and reduce mass-transfer inefficiencies, but they limit overall flow rate. The smaller plants can also be sited in areas unable to accommodate large-scale industrial facilities.

Small GTL plants can be built close to isolated sources of excess methane, stranded gas. Landfill gas—primarily methane and carbon dioxide—is one example of a typical stranded gas; another is associated natural gas produced in oil fields that have little or no natural gas infrastructure. GTL plants in such places could potentially obtain feed gas at steep discounts or even for free, since stranded gas is usually flared, burned off or vented, allowed to dissipate into the atmosphere. Small-scale GTL plants could become a more attractive option than flaring in the future, depending on the finalized version of rules initially proposed in February 2016 by the Bureau of Land Management designed to limit the amount of methane flared or vented from oil and natural gas production activities. If this gas were converted to liquid instead, it could be transported by vehicle or pipeline and sold.

IMO Fuel Sulfur Limits on Maritime Transportation Will Spark Changes by Both Refiners And Vessel Operators

The International Maritime Organization IMO, the 171-member state United Nations agency that sets standards for marine fuels, decided in October to move forward with a plan to reduce the maximum amount of sulfur and other pollutants present in marine fuels used on the open seas from 3.5per cent by weight to 0.5 per cent by weight by 2020. This decision follows several other marine fuel regulations limiting sulfur content, such as the implementation of Emissions Control Area ECA requirements in coastal waters and specific sea-lanes in North America and Europe, where the maximum

sulfur content of fuels was limited to 0.1per cent by weight starting in July 2015.

The sulfur content of transportation fuels has been declining for many years due to increasingly stringent regulations. In the United States, federal and state regulations limit the amount of sulfur present in motor gasoline, diesel fuel, and heating oil. New international regulations limiting sulfur in fuels for ocean-going vessels, set to take effect in 2020, have further implications for both refiners and vessel operators at a time of high uncertainty in future crude oil prices, which will be a major

factor in their decisions.

Bunker fuel—the fuel typically used in large ocean-going vessels—is a mixture of petroleum-based oils. Residual oil—the long-chain hydrocarbons remaining after lighter and shorter hydrocarbon fractions such as gasoline and diesel have been separated from crude oil—currently makes up the largest component of bunker fuel. The sulfur content of crude oil tends to be more concentrated in heavier hydrocarbon molecules, with heavier petroleum products such as residual oil having higher sulfur content than lighter ones like gasoline and diesel.

December 2016 • Issue 12

Additionally, the state of California and the European Union have regulations on the sulfur content of marine fuels, and the types of fuel used when ships are at dock, waiting to dock, or are maneuvering within port. For example, a vessel approaching the port of San Francisco may have to change its fuel mix twice: once when going from the open seas higher-sulfur fuel of mostly residual oil, to an ECA compliant lower-sulfur fuel mix, and again to a marine diesel fuel compliant with California's ocean-going vessel regulations for use within ports.

The IMO sulfur limits that take effect in 2020 will affect the fuel used in the open seas, the largest portion of the approximately 3.9 million barrels per day of global marine fuel use, according to the International Energy Agency, presenting several challenges for both refiners and shippers.

The first challenge for refiners is to increase the supply of lower-sulfur blend stocks to the bunker fuel market. Refiners have several potential paths. One approach is to divert more low sulfur distillates into the bunker fuel market. Another option would be to use low sulfur intermediate refinery feedstocks in bunker blends. In both cases, care is required to assure that new fuels continue to meet specifications for use in marine engines.

A second challenge for refiners is what to do with the high sulfur residual oil that can no longer be blended into bunker fuel. Adding capacity to desulfurize residual oil is one option, but the economics do not currently appear to be attractive. An alternative strategy is to build or expand refinery units that take heavy hydrocarbons, such as residual oil, and upgrade them into lighter, more valuable products, but this would require large investments. In either of these cases, refineries would be faced with investments and costs that are acceptable only if there is certainty of future demand from the shipping industry.

Vessel operators also have several choices for compliance with the new IMO sulfur limits. For example, IMO regulations allow for the installation of scrubbers, which remove pollutants from ships exhaust, allowing them to continue to use higher-sulfur fuels. Some ship owners that operate primarily in coastal areas, such as cruise lines and ferries, opted to install scrubbers on their vessels as the new ECA regulations came into force. The possibility of widespread scrubber installations, which would allow for continued use of higher sulfur residual oils, could make refiners hesitant about making large investments to build refining units capable of upgrading the residual oils.

Ships also have the option of switching to new lower sulfur blends or to non-petroleum based fuels. Some newer ships and some currently being built have engines that would allow them to use liquefied natural gas LNG rather than petroleum-based products. However, the infrastructure to support use of LNG as a shipping fuel is currently limited in both scale and availability.

Vessel operators and shippers will also likely be faced with the higher costs as the sulfur content in marine fuels decreases and the role of distillate in the bunker fuel market increases. An example of the price difference between fuels can be observed at the refining and trading hub in Northwest Europe, known as the ARA, collectively the cities Amsterdam and Rotterdam, in the Netherlands and Antwerp, in Belgium. Prices for low sulfur gasoil, a type of distillate, in the ARA has averaged over $20 per barrel more than high sulfur fuel oil, residual oil for use as a fuel, to date in 2016. Fuel blends used to meet the new IMO regulations are likely to price somewhere in between these two fuels.

IEA moves to Enhance Global Gas Security

While the rise of the liquefied natural gas market has accelerated the globalization of natural gas, the energy security implications of this transformation have attracted much less attention. Through an extensive analysis of global gas data, a new report from the International Energy Agency seeks to provide more transparency into the LNG market.

There is no doubt that global gas markets are well supplied today. While this is positive for global gas security, the new analysis from the first Global Gas Security Review, released in Tokyo, warns that LNG markets are less flexible than is commonly believed.

A growing share of LNG capacity is offline – mostly because of a lack of enough gas to feed into the system but also because of security and technical problems – meaning the market has less extra capacity than assumed. Between 2011 and 2016, the level of unusable export capacity has doubled, disabling about 65 bcm of gas, which is equal to the combined exports of Malaysia and Indonesia, the world's third- and fifth-largest exporters. A period of low oil and gas prices could further worsen the situation.

However, the Global Gas Security Review finds that LNG contract structures are becoming less rigid, increasing market liquidity. In 2015, about 40 per cent of LNG contracts had fixed destination terms, down from 60 per cent for contracts signed up to the year 2014.

While shorter term contracts are gradually becoming more common, buyers are also accepting longer contracts in exchange for increased flexibility in the final destination in order to better respond to market conditions. **Flexible contractual structures are important for gas security as they enable to aggregate gas volumes at a lower cost from various regions.**

LNG's share of the global gas market is set to increase in the coming years. In fact, LNG supplies have grown at a faster pace than total gas consumption. Providing a factual picture and analyzing its implications for gas security matches well with the IEA's core mandate about energy security.

"The growth in the global gas trade, along with the diversification of supply sources, is improving the security of supply," said Fatih Birol, the executive director of the International Energy Agency. "But there is still a need to be vigilant on gas security as the changing nature of the market means that regional demand and supply shocks may now be felt in more distant places than ever before."

The report provides detailed case studies on Europe and Japan. For Japan, it shows that while gas markets reacted relatively effectively to the loss of nuclear generation in Japan after the Fukushima nuclear accident, the factors that made that possible cannot always be counted on in the future.

The Global Gas Security Review builds on an extensive set of data and other substantial inputs from industry and will be produced on an annual basis. It is accompanied by country specific statistics; data on outages of LNG export capacity by type and region; flexible LNG demand by importers; flexible LNG supply by importers, producers and portfolio players; and flexible gas demand and supply in Europe.